FIRE ENGINES OF YESTERDAY

by

JOHN CREIGHTON

IAN HENRY PUBLICATIONS
1984

187245

ISBN 0 86025 878 5

Printed and bound in Great Britain by
Biddles Ltd, Guildford and King's Lynn
for Ian Henry Publications Ltd,
38 Parkstone Avenue, Hornchurch, Essex RM11 3LW

CONTENTS

Most of the illustrations come from the
John Creighton collection.

Thanks are extended to all who provided material
for this book, especially
Malcolm Tovey, R Adelman, M Boucher, Don Feipel,
Danie Imhof, B.S.P.P., Musee des S. P. de la Communaute Urbaine de Lyon.

This pump escape is an ATC/Merryweather Marquis that featured heavily in
Merryweather advertising. Based at Southend-on-Sea, it finished up as a spare
appliance, its last regular station being Leigh.

United Kingdom

The mood of the moment is captured as the crews set about connecting up lengths of suction in 1884. Shand Mason had built their first steam engine for operational use in 1858, while 1861 saw Merryweather's "Deluge" come on the scene featuring a single horizontal cylinder and double acting pump.

Built about 1880, this fire pump is a reciprocating design capable of discharging 500 gallons of water a minute, and this Shand Mason steam fire appliance has a multi-tubular boiler allowing steam to be raised from cold to 60 p.s.i. in about eleven minutes.

Bowling Street Fire Station, Leicester, 1892. A remarkable view of two horse-drawn steam-powered fire engines. The Superintendent is the third from the left in the front row. The complement of the Borough Brigade about this time was 10 permanent and 21 auxiliary firemen.

A contemporary shot of Leicester's Rutland Street Fire Station and horse-drawn appliances, circa 1849. The station had cottages for firemen at the rear of the building, each provided with an electric fire alarm bell, and records from the mid-1800s indicate that horse-drawn manual appliances travelled up to 8 miles in order to deal with fires.

Wymondham Fire Brigade was established in 1882, the Norfolk town acquiring two Shand Mason manual pumps at first, including this 22-man 'manual' capable of throwing 100 gallons of water a minute to a height of 120 feet. The appliance was towed to jobs by cab horses hired from a local business, this continuing until 1922, when a hired lorry replaced horses. Volunteer men were summoned to the station by maroons set off by the police.

Pictured outside London Road Fire Station, Edinburgh, in 1902, this horse-drawn pump carried a large bell under the driver's foot-rest. It was about this time that the Firemaster, Mr Pordage, persuaded the Finance Committee to purchase 7 additional horses and to increase the number of steamers in the city to 6.

Horse drawn appliances were quite popular at the start of the 1900s, this one carrying an extending ladder. The crews wore sailor-like caps in some areas and qualities in handling horses were looked for in new recruits.

A splendid photograph of a steamer turning out during 1903. When these appliances were called to an incident the engine boiler was lit before leaving the station and steam was raised in about eight minutes while en route to the job.

The year is 1912 as London Salvage Corps line up outside their Watling Street premises. The HQ was established at No.31, with 3 out-stations in the West, East and South-East of London. In 1905 the Corps transferred to 63/69 Watling Street, staying there until 1961.

This period shot is of a 1903 Motor Fire Brigade Chemical Tender capable of travelling at 18-20 m.p.h. Such a machine would have cost £650 and the 4-cylinder, 24 h.p. model had a copper cylinder, containing a chemical charge of 50 gallons, and a 180 foot hose reel, plus a box with 150 yards of canvas hose.

Manned by fire personnel, this 1912 motorised ambulance took 3 patients, driver and attendant. Costing about £590, the Siddeley/Deasy 18/24 h.p. model had bodywork by Midland Counties Motor Engineers.

One of the first limousine pumps in service with Edinburgh Fire Brigade, the 1932 Dennis was the 'Big G' model, this marque also seen in other forms, such as pump escapes. Pictured in NFS livery during World War II, the machine has white paint running the length of the body and on the front mudguards, to help during black-outs when street lighting was non-existent. The Dennis carried ladders on the roof and inside.

An interesting view of the 1937 Morris Commercial in service with London Salvage Corps until 1952, carrying a variety of gear including waterproof salvage sheets, drain clearing equipment, brooms and shovels. Morris Commercials could be seen with several authorities, including some Austrian brigades, in the early 50s, and in airfield tenders based on the six-wheeled chassis.

The 1935 Dennis 'Ace' was the first motorised appliance at Wymondham and, although this model was an open type, the mid-30s saw several brigades having enclosed fire appliances based on 'Ace' chassis. The Wymondham machine could pump 350/500 g.p.m. It remained on the run until 1956.

Dennis Patent Turbine Fire Engine and Escape Ladder

Built in 1938 with Merryweather turntable ladders, this machine comes with a Merryweather chassis. The company had made its mark some five years before the Barking Fire Brigade TL was built, when they produced the first all-British 100 foot motor turntable ladder with steel sections, delivered to Hong Kong in 1933.

Seen in action with bonnet open, the 1938 Dennis Light Six Limousine Pump was purchased as a pump, but later employed as a Pump/Engineering Tender. This particular machine is unusual as a limousine appliance in that it has a rear-mounted Dennis Humber pump.

GYR 724 is a Bedford 'QL' ex-Army Fire Service Tender; the only civilian modification was the lowering of of tilt height and removal of ladder. The vehicle is towing a Coventry Climax Medium Trailer Pump and, when operational, this appliance was based at Rayleigh, Essex.

A good example of a wartime fire appliance, a 1939 Bedford operated by the Rover Motor Company Fire Brigade seeing action during the Coventry blitz. The insignia on the door shows that it was based at Solihull; the pump was positioned behind enclosed crew quarters.

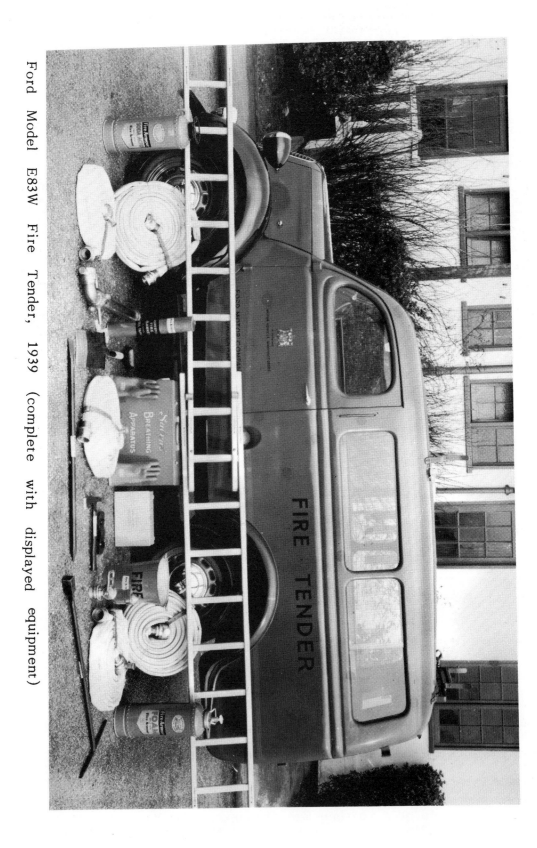

Ford Model E83W Fire Tender, 1939 (complete with displayed equipment)

The Fordson 7V Escape Carrying Unit was a familiar sight in the 1940s; GGK 121 was built in 1940 and started work with the Home Office in London. The appliance had a Barton pump and was powered by a Ford V8 side valve 3.6 litre engine. Seen in its war-time grey livery, the vehicle has masked headlights, while the pump and hose reels were provided by the NFS in 1944 before the machine was moved to Horsham Fire Station, West Sussex, where it served as a spare Escape Carrier until 1961. It was acquired for preservation in 1981.

Austin Standard Towing Unit with Coventry Climax Trailer Pump, issued in 1941 for war-time firefighting, and pictured among blitz damage. The Auxiliary Towing Vehicle was one of the first machines of its type ordered by the Government in 1941, specially for towing trailer pumps.

Salvagemen from Liverpool at an incident in the 1940s with their vehicle behind them. Appliances carried sheets, ladders and brooms, etc., and up to the end of the Salvage Corps in 1984, the modern machines were equipped with sophisticated salvage gear, plus breathing apparatus sets, two-way radios, etc. The Liverpool Corps gave assistance to fire operations and reconditioned 26,000 bales of cotton from warehouses and 20,000 from ships during the war.

Pictured during World War II, the Ford WOT 2V8 fire appliance stands in front of bombed premises. Used by the National Fire Service, these machines had a crew of 8 and towed a trailer pump. These vehicles were based on the Ford Model 61 and E88W, with modified running gear and chassis.

A fine shot of a Fordson 7V water tender equipped with a Ford V8 30 h.p. engine. This is a good example of a postwar rebuild of a wartime utility tender – note the early postwar accent on foam carriage, originally used as water carriers. Postwar modifications include hose-reel fitting and semi-permanent 'plumbing-in' of the portable pump.

A 1954 Bedford CA of Kesteven Fire Brigade, later employed at Sleaford as a hydrant and general purpose van until 1974. HTL 913 then moved to Lindsey County Council for a further 16 months.

One of 3 machines built in 1958 by HCB-Angus for the North Riding of Yorkshire County Fire Brigade, this Bedford is based on a TD4 chassis with a 300 cu.in. 6 cylinder petrol engine, carrying a 35 foot Ajax ladder. This machine was one of the first to be equipped with HCB's own power take off unit, enabling her to pump on the run. SAJ 806 served her time at Lythe station, later moving to Richmond as a spare. Taken out of service in 1979.

Germany

A pleasing example of early international fire appliance manufacturing is found in this 1906 vehicle, whose chassis came from the Salford firm of John Morris, other components by Magirus. The ladder was a type 'DK' and would reach a height of 26 metres.

The first motorised fire engine, Gaggenau (Baden), dating from 1907 and weighing 2,430 kg., with solid rubber rear wheels and pneumatic tyres at the front. This early example of a Mercedes vehicle could achieve speeds of up to 35 kilometres an hour and its exposed water tank was mounted at the rear.

The Mercedes Company provided its own fire protection in the shape of this appliance dating from the early 1900s. In common with many German and French brigades, the machine had a removeable hosedrum mounted at the rear; the pump delivered 1,500 litres of water per minute.

Two early Mercedes fire engines, the one on the left appearing for the first time in 1907 (D4), while the other is a type E4 of 1909. Both machines are right-hand drive models, featuring reinforced rear wheels designed to take the weight at the back of the machines.

The Opel factory in Germany, 1912, where the pumping capabilities of a motorised fire appliance are tested. Water is drawn through suction from an underground tank, and much of the equipment is carried on the outside of the appliance.

Carrying a crew of 11 and powered by a 4 cylinder engine, this Daimler-Benz fire engine was built in 1926 with a right-hand drive and 4 x 1 gears. The pump was capable of delivering 1500/2000 litres a minute.

A 1926 Magirus 3CS carrying lengths of suction hose on the outside of the appliance. The length of hose was added at the rear on delivery to the brigade.

On a Magirus 3CL chassis, this 30 metre turntable ladder was built in 1926 and saw considerable service with the Magirus Factory Fire Brigade at Ulm.

A 1928 Mercedes-Benz appliance whose fire fighting facilities included a large water tank, together with foam and high pressure pump. Receiving power from a 6 cylinder engine, the machine could pump 1,500 litres of water a minute and had left-hand drive.

The wholetime firemen of Stuttgart pictured in their 1928 Mercedes
fire appliance powered by a 68.PS. engine and carrying a pump
delivering 2,000 litres a minute.

The Metz logo can be seen on the 26 metre turntable ladder dating from 1929, the first all-metal Metz ladder appearing five years earlier. The appliance shown had brakes acting on all four wheels, could carry 32 kg of petrol and was driven by a 6 cylinder engine.

Metz received a patent for turntable ladders with hydraulic gear during 1935, when a 45 metre ladder was delivered to Kingston-on-Hull, U.K. The turntable ladder shown was manufactured the following year, comprising a 45 metre, all steel, 6 stage unit, driven by a 6 cylinder engine.

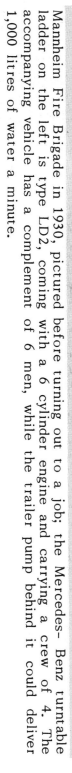

Mannheim Fire Brigade in 1930, pictured before turning out to a job; the Mercedes-Benz turntable ladder on the left is type LD2, coming with a 6 cylinder engine and carrying a crew of 4. The accompanying vehicle has a complement of 6 men, while the trailer pump behind it could deliver 1,000 litres of water a minute.

This Opel is in prime condition having just left the factory, its freshly-painted Bachert pump bearing the company name. It is still possible to find front positioned pumps in appliances in many European brigades.

This 1930 Opel 'Blitz' carried its name on the engine cowling and featured a bell for audible warning, together with a roof-mounted spare tyre. During the Second World War, Opel produced several appliances such as the 1943 Blitz 3.6 - a combined water tender and pump unit.

Metz bodywork is featured on this 1949 Opel Blitz that has front mounted pump and enclosed seating for the crew.

Some 40 years after its first appearance, this 1944 Mercedes-Benz L 4500 is still on the run with Stuttgart Fire Brigade. Used as a foam tender, the vehicle receives power from a 120 h.p. diesel engine and carries a 3,000 litre tank for fire fighting purposes.

The Opel 'Blitz' LF8 (1952) with Metz bodywork has a front mounted pump, plus two flashing lights above the cab.

A Mercedes-Benz DL 37 turntable ladder with a lift for rescue purposes, produced between 1950 and 1958, Metz introducing their all-hydraulic ladder in 1958.

Austria

Built in 1913, the horse-drawn appliance came with Rosenbauer and Fiat components. The motor was a C2 model with 26 h.p. and the machine had a water output of 500 litres at 6 bar. The Rosenbauer Company began in Austria in 1866 and in 1908 the first motor driven centrifugal pump came on the scene.

An interesting line-up at Graz of fire appliances built on WAF chassis (Vienna) between 1914 and 1918.

Dating from the early 1930s, this portable pump, type R50, was popular with many Austrian brigades, powered by a two-stroke two cylinder engine, delivering 600 litres of water at 10 bar.

This 3 axle appliance was on the run with Vienna Fire Brigade in 1934, built on the Austro Daimler cross country chassis. Some U.K. and German brigades had electric turntable ladders built by Austro Daimler and marketed by Cedes Electric Traction, Ltd. The appliance illustrated is a multipurpose fire tender carrying a high pressure pump (30 bar).

France

This 'voiture electrique' made its debut in Paris during 1889, powered by 650 kg of batteries, producing a speed of 19 kilometres an hour. There were 3 such vehicles in the city, each carrying a crew of 6 and providing water for 3 jets, but the recharging of batteries proved something of a problem and the machines were only useful on relatively short journeys.

Senior officers of the Paris Fire Brigade would travel in this 'Voiture Automobile' in the early 1900s. The Bataillon de Sapeurs-Pompiers de Paris was created in 1811, comprising five companies in 1840, receiving in 1904 the first motorised 'steamer'.

An interesting 'Steamer' from Masson dating from 1893, when it was in service with Lyons Fire Brigade as a horse-drawn vehicle; the shafts have been removed and two brass branches are positioned near the driver's seat.

Hard at work with a 1908 'Traction Hippomobile', these French firemen are preparing to pump water. One problem facing fire personnel arriving at an incident was that the horse had to be removed from the shafts to gain access to the pump.

Several French fire brigades had long wheelbase personnel carriers in the early 1900s, capable of transporting 15 men to an incident. This one saw service with the Roubaix Fire Brigade and was known as 'le chariot de corvee'.

Lille had several examples of 'steamers' on the run, the first acquired in 1876, supplied by Shand & Company, London. The machine illustrated was operational in 1900 and carried lengths of suction hose along the side of the 'pompe a vapeur'.

2 horses were needed to pull this heavy machine whose complement was a dozen men. The stowage of ladders sometimes created problems for firemen using this appliance in the early 1900s.

This 'Autopompe' from Roubaix can take 10 men. In 1701 the areas of Roubaix and Tourcoing signed an agreement for mutual assistance at times of fire.

Roubaix firemen use a 24 metre ladder to scale this building during an exercise.

The year is 1903 - Lille Fire Brigade has a 'shout' and four horse-drawn vehicles are ready to depart for a fire. Today the City's Brigade responds to some 35,000 incidents a year, the manning provided by 600 volunteer and 800 paid firemen.

The English firm, Merryweather, sends its products all over the world and this turntable ladder was built in 1908 and acquired by Lyon Fire Brigade during 1923. With a working height of 26 metres, the TL had accommodation for a crew of 2 and, although operational in France, had right-hand drive.

A 1910 'Auto-pompe' powered by a Vermorel engine (10 CV, 4 cylinders). The appliance had 4 speeds and a centrifugal pump; lengths of hose were carried at the rear.

A good example of a 1914 Mieusset 'Auto-Pompe' equipped with a 4 cylinder engine and pump from Mieusset, the latter discharging 1,000 litres of water a minute. Lengths of suction are carried at the rear, while the length of hose is on a drum ready for immediate use.

A Rochet Schneider (RS) fire engine from 1924, powered by an 18 cu. four cylinder engine alongside a 1932 Citroen that was in service with firemen in the Isere region. A red flashing light was positioned on a horizontal bar between the headlamps.

A Somua engine and pump came with this 1926 fire appliance with power from a 4 cylinder unit. Carrying a crew of 8, this machine had braking on all four wheels.

38

Delahaye First Attendance vehicles were a common sight in France for many years. This 'Premier Secours' machine first made its appearance in 1936 (Type 140-103) and is equipped with a 6 cylinder petrol engine (90 cu.). Carrying a 400 litre water tank, the vehicle delivers water by means of a centrifugal 'Rateau' pump.

With a crew of 6 and centrally mounted hose reel, the Laffey 'Premier Secours' appliance was first put into use during 1939; the type B SSC 3 vehicle equipped with a Laffey centrifugal pump, plus a 500 litre water tank.

French fire brigades often have examples of vehicles dating from the Second World War period. The 1941 'Wrecker' illustrated comes on a three axle chassis and this Ward France heavy crane has both front and rear mounted winches – 9 tons and 22 tons capacity respectively. The crane at the rear is useful at incidents where debris has to be cleared away, and this particular vehicle was in service with the French army before going to the fire brigade in the Moselle region.

A quarter century old, the 'EM.20' has Citroen chassis and bodywork and is still operational in Thionville. The vehicle carries a 20 metre ladder and is powered by a petrol engine.

Switzerland

The first permanent fire station in Geneva had 4 men who took up duties in 1899 and this horse-drawn steamer was one of the first appliances used in the city. Delivering 1,200 litres of water per minute at eight bars, the vehicle weighed 1160 kg and it took a quarter of an hour to put the pumping mechanism into operation.

Dating from 1883, this Merryweather Steamer was in use until 1904, pumping 1,000 - 2,000 litres of water a minute.

A remarkable three-man apparatus seen with Basel Fire Brigade first in 1899, constructed by Julius Dressler. The third man carries a stirrup pump while his colleagues come equipped with personal rescue lines and axes.

One of the early motorised fire appliances in Geneva, going on the run in the early 1900s and carrying a length of hose at the rear. It was about this time that the system of 24 hours on duty and 24 hours free was introduced and by 1914 Genevan firemen were regarded as civil servants.

Chassis from Lohner-Porsche & Co., Vienna, and electric motor from System Lohner-Porsche - a remarkable electric powered fire appliance used in Switzerland from 1905 to 1931. In spite of impressive looks, this machine was limited to fairly short journeys owing to the need to recharge batteries.

The fire fighting scene at the start of the 20th Century is captured in this fine example of a 'steamer' used by Basel Fire Brigade from 1905 to 1925. Delivering 2,100 litres a minute, the motorised appliance is pictured just after delivery to the Brigade, the officer in charge wearing a cap, travelling at the front.

Shown outside Geneva's Fire Brigade HQ in Rue du Soleil-Levant in 1917, the Picard-Pictet fire engine was known as the 'Pic-Pic', one of the first motorised appliances in Switzerland. The letter 'F' denotes 'Feu' and the machine is actually a converted taxi, used by the 10 firemen, the complement rising to 19 during 1922.

The Swiss firm of Saurer made the chassis for this machine first put to work during 1931 and remaining in service until 1974. Equipment included a centrifugal pump delivering 1500/3000 litres a minute. Crew members sat in the front seat and alongside the machine, while a form of wheeled escape featured among the ladders.

One way of arriving at a fire! Swiss firemen pictured in 1931 ready to move off to provide extra personnel at an incident on their Motosacoche SA.

Geneva has always boasted a fascinating collection of fire appliances, as shown in this view of the city's third permanent station, opened in 1932 and replaced in 1957 by a more modern building.

Basle Fire Brigade acquired St Florian in 1939, the vessel remaining on station until 1973. The ship was powered by two Sulzer-Schiffs diesel engines (2 x 125 PS) and carried two horizontal centrifugal pumps, the output being some 7200 litres a minute. Today there are two fire-boats based at Basle - 'Basel-Land' and 'Basel-Stadt', employed for ice-breaking, water rescues, oil pollution and fire-fighting.

Holland

During the late 1600s, two Dutch brothers, Jan and Nicolaas van der Heijden, introduced leather hose for delivery and suction, together with couplings for joining lengths. It was not until the 1800s that copper rivets replaced sewing along the seam. In 1870 a major fire destroyed many areas of Amsterdam; the city's wholetime fire brigade became operational 4 years later. The first 3 pictures show Dutch firefighting apparatus employed in the latter half of the 19th Century and the early 1900s.

Built in 1959, this Dutch machine had a Magirus Deutz F6L613 chassis and Kronenburg body. Carrying a 2000 litre water tank, the vehicle saw service in Dordrecht until 1980 when, in May of that year, it was sent to the Central American state of Honduras where it is on operational duties.

Scandinavia

The year is 1888, the machine is a horse-drawn manual fire engine on the run with Copenhagen Fire Brigade. Each fireman has a safety line for use in the event of a rescue and the rear positioned length of hose is readily accessible.

A 'Triangel' chassis forms the basis of this Danish pump built in 1936, equipped with bodywork from H Meisner-Jensen. One interesting feature is the exposed rear mounted pump and lengths of hose stowed at the side and rear of the appliance. The Aster pump delivered 2000 litres a minute and the vehicle was powered by a Triangel 100 h.p. engine.

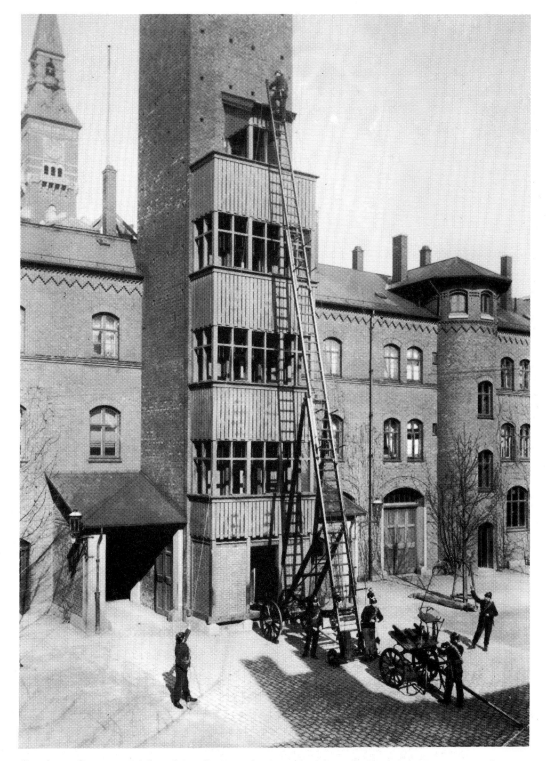

Dating from 1886, this horse-drawn ladder belonged to Copenhagen.
The chassis/ladder came from Hassel & Tendt, the working height
being 19 metres. This period view shows the equipment in use in the
1890s at Copenhagen's Fire Brigade HQ.

Gothenburg relied on Swedish-built fire engines in the 1940s, this LV 102 rescue crane appearing in the mid-40s, seen here demonstrating its lifting capacity. Other instances of Volvo fire appliances with Gothenburg included an LV 192 turntable ladder (1928) and LV 81/94 fire vehicles.

The popular Volvo LV 80/90 series of fire vehicle made its appearance in Sweden in the late 30s, carrying a hydrant standpipe on the running board, while a rear mounted pump was positioned next to a first aid hose reel.

A Volvo LV 94 fire appliance, popular with many Swedish brigades during the late 1930s and 40s. Carrying a comprehensive range of firefighting equipment, it had a useful wheeled escape and front mounted pump. Volvo Commercial Trucks were built on the LV 94, these followed by the LV 180, 190 and 290 series in 1937.

A neglected Volvo pumper, no longer in use and left to the elements in a Swedish garage - one interesting detail is the location of a hose reel adjacent to the front mounted pump.

Oslo Fire Brigade had this 1936 Magirus turntable ladder mounted on a Magirus chassis. With a working height of 37 metres, this machine was powered by a petrol engine and was operational until 1963.

Fordson fire appliances could be found with many U.K. brigades in the 1940s, often employed as utility escape tenders, and so it is interesting to discover a 1932 Fordson still on the run with Reykjavik Fire Brigade. Carrying a crew of 2, this open appliance is in use alongside a variety of vehicles, including ex-RAF 'Green Goddess' pumps and USA Fords supplied by Darley of Chicago.

On the run in 1948 at Oslo's Station Eight, this Ford machine came equipped with a front-mounted Barton pump.

Received by the Oslo Brigade in 1955, the 'Ford Big Job Water Tender' carried a Barton U-50 pump positioned at the front of the machine with a tank holding 1200 litres of water; this fire engine had a crew of 5 and had distinctive two-tone livery.

Built in 1963 and in service with Oslo Fire Brigade, this 37 metre Magirus turntable ladder is a 175 D. The machine is now out of service and was the former engine 93.

Built to withstand northern winters, a WIBE turntable ladder is on a Dodge chassis, receiving power from a petrol engine. Some 20 years old, this Baerum Fire Brigade appliance displays an 85 foot ladder and ample locker space.

Czechoslovakia

Made in 1945, the Skoda 256 is a rare example of a preserved Eastern Block fire engine of this type and now belongs to a fire enthusiasts' group. The Czech firm of Skoda has produced a variety of fire appliance, including the ASC of the 1960s, while Prague Fire Brigade had examples of the Skoda 1203 with a portable pump and crew of 5.

There are very few Praga A150 fire engines in use today; this one was built in 1948 and weighs 1½ tons. Powered by a 55 b.h.p. four cylinder petrol engine, the Praga A150 was employed in parts of Austria, in addition to Eastern Europe.

First operational in 1951, this Praga RN receives power from a petrol engine and, in addition to acting as pumping machines, RN models could be found as turntable ladder appliances with Metz ladders.

One of the most popular fire trucks in the 1950s in Czechoslovakia, this Praga RND came with a diesel engine and weighed 3 tons, the pump being mounted at the front. A more recent Praga A-30 machine is equipped with Karosa components designed as an aircrash tender, carrying some 34 carbon dioxide cylinders.

Built in 1954, this example of a Praga RN featured a 1600 litre water tank, front mounted pump and lengths of suction carried on the running board. Praga also produced 6 wheel chassis seen with brigades, with army units and for civilian purposes.

South Africa

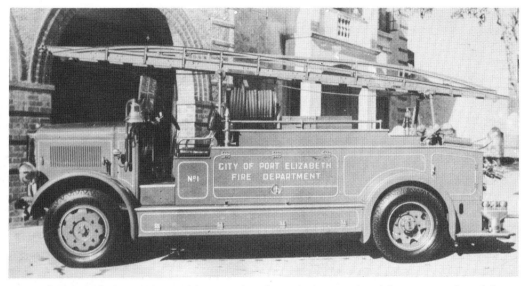

Manufactured in 1934, this Leyland Cub Pump/Ladder served with Port Elizabeth Brigade and is now in a playground.

Another Leyland Cub Pump/Ladder dating from 1934 and operational with East London Brigade. When taken out of service it was restored by Brigade members and a number of parts were taken from the Port Elizabeth vehicle.

Australia

A 1942 Canadian Ford Blitz is the basis of this bush fire tender, originally an aircrash tender with the Royal Australian Air Force during World War II. The tank is the original 500 gallon fixture with a pump motor taken from a 1950 Ford 100E Sedan. The 'Blitz' was purchased at an Air Force auction in 1950 for about £200 and, such is the pride in this 40 year old machine, that an inspection of the vehicles reveals that even chassis belts are polished!

A 3 axle 1942 Studebaker still in service, the body comprising several purpose-built specifications added to improve the performance of a war-time fire engine. Carrying an 85 g.p.m. Tankermaster pump, the Studebaker has a 750 gallon water tank.

A story surrounds this 1942 Studebaker, which was to be part of a consignment of 19 bound for Russia via the US Navy: the Russians refused them, so they were landed in Australia. Used by the country's army as a radio unit, the vehicle was bought in 1975 with 5 similar machines at a total cost of A$2,500. Carrying a 600 gallon tank and a 1948 Coventry Climax pump delivering 550 g.p.m., the machine saw most of its action with bush fires.

Woodstock is some 250 miles from Sydney and its Blitz Buggy dates from 1942, carrying a 500 gallon water tank and 150 g.p.m. Davey pump. During its time as a fire engine, this vehicle has been scorched in raging bush fires and has been rebuilt several times.

Restored by a fireman in Lilydale, Victoria, the 1927 Graham Brother vehicle has a Dodge Flying 4 engine, magneto ignition and three speed gearbox. Tyres are 20 x 5 8 ply. Restoration took 3 years. About 30 Graham Brothers fire engines were constructed, but only 5 are still inexistence in Australia.

The New South Wales Fire Brigades owned the 1926 Garford Type 15 Pumper, carrying a short ladder, water tank and first aid fire fighting equipment. There is provision for 2 crew members in the front seat, while others sat at the back on either side of the ladders.

Carrying a crew of 8, the Dennis 'Big 4' appliance came on the scene during the late 1930s. Pumping 600 gallons a minute, this vehicle had an exposed rear mounted pump, ladders and lengths of suction stowed under the feet of rear seated firemen.

During the 1940s Mack fire vehicles could be found in several areas of Australia; the appliance on the left is a 1942 Salvage Tender, a Pumper is next, while from the same era, a Mack Turntable Ladder Machine carries an interesting 1929 Magirus Ladder.

United States of America

The name Oshkosh comes from the Indian Chief of the Menominee tribe that lived in the area. Fire buffs will doubt-less associate the town with the Oshkosh Truck Corporation founded by N Besserdich, initially making a variety of com-mercial vehicles, the first fire engines appearing in 1920. The illustrations show some of the first Oshkosh fire vehicles in service with Oshkosh Fire Department, which favoured 4 wheel drive engines carrying ladders along the side of the appliance.

A 1932 Seagrave fire appliance on the run with Laramie County, Wyoming. Frederick Seagrave started a company in 1887 that built ladders used for fruit harvesting in Michigan, but started constructing fire appliances at the turn of the century. In 1912 he developed a centrifugal pump more powerful than any rival.

The town of East Brunswick, New Jersey, had a volunteer unit in the mid 30s and among their 5 engines was this 1936 Diamond - 'T'. With bodywork by the General Fire Truck Corporation the vehicle had a Waterous pump delivering 500 g.p.m., while a first aid hose reel was located behind the driver.

The South Old Bridge Volunteers, N J, owned the 1937 Salvage Truck with a Dodge chassis and bodywork from Approved Fire Equipment Company. Audible warning was supplied by a siren on the cab roof, while two flashing lights were positioned on the mudguards as visible warning devices.

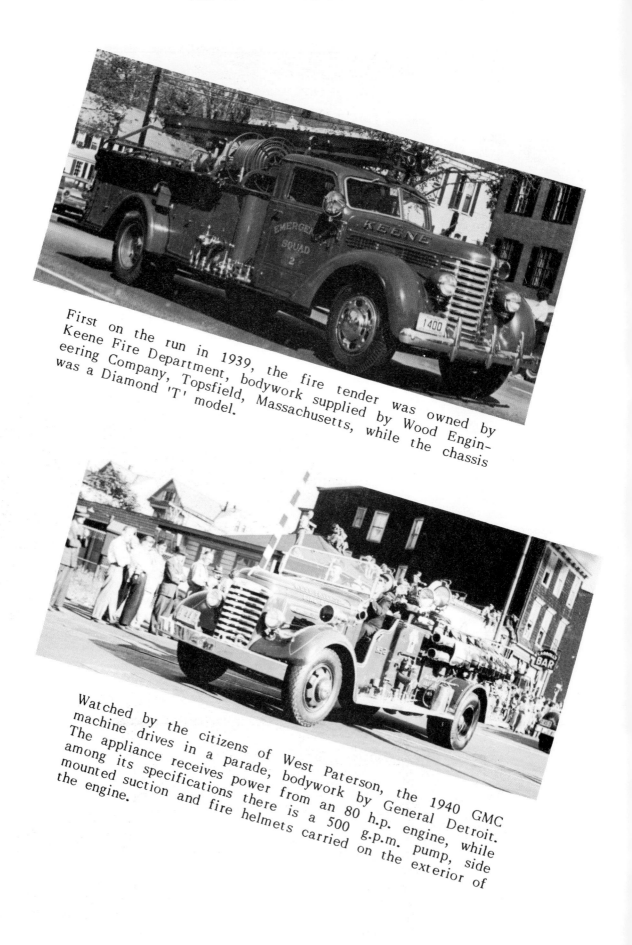

First on the run in 1939, the fire tender was owned by Keene Fire Department, bodywork supplied by Wood Engineering Company, Topsfield, Massachusetts, while the chassis was a Diamond 'T' model.

Watched by the citizens of West Paterson, the 1940 GMC machine drives in a parade, bodywork by General Detroit. The appliance receives power from an 80 h.p. engine, while among its specifications there is a 500 g.p.m. pump, side mounted suction and fire helmets carried on the exterior of the engine.

Pictured in prime condition, a Cheyenne 1943 Seagrave awaits a fire call. The flashing light is in an elevated position, while the rear mounted spotlights give illumination at night incidents.

A robust looking Rescue Squad vehicle with Hamilton Fire Company was called to incidents, such as road traffic accidents or rail crashes. Incidents after dark were aided by the many searchlights carried on this 1946 Dodge, while a front mounted winch was an added advantage at rescues.

The Peter Pirsch Company was started in 1900 and 1926 saw the first Pirsch built motor chassis. Dating from 1946, this Pirsch pumper was on the run with Oak Forest F D, Illinois, having a flashing lamp on the cab roof and at the front of the bonnet. The vehicle carried both a siren and bell for warning traffic to move aside. The Pirsch group had many 'firsts' in fire appliance manufacture, including the first all-aluminium aerial machine.

Termed the 'Quad' fire truck, this 1955 Ward LaFrance vehicle had four principal features in the pump, water tank, hose body, 204 foot ground ladders. On the run in Wharton, New Jersey, the pumper delivered 750 gallons of water a minute at 150 lb.p.s.i. and was driven by a Waukesha 6 cylinder petrol engine. The pump is of the Waterous 2 stage variety.

A powerful looking Seagrave pumper with Los Angeles City Fire Department in the 1940s, Engine 38 had a pump delivering 1000 g.p.m., monitor located behind the driver, midships pump and flaked length of hose ready for use on the running board.

Blairstown Fire Department, New Jersey, had its own Ambulance Corps manning this 1951 Dodge Emergency Truck carrying custom-built coachwork and a variety of rescue gear, including a front mounted winch. Personnel were trained in both ambulance and rescue work.

First operational during 1954, the Seagrave aerial appliance carried a
65 foot ladder and provided fire cover at Elmhurst, Illinois.

Industrial premises and oil refineries often have their own fire apparatus, as shown by Union Oil Refinery, Lockport, whose 1954 Darley/
Ford (500/300) carried a midships mounted pump.

Jeep vehicles can be found throughout the world with a variety of bodywork, including Rosenbauer and Willys. The vehicle illustrated comes with coachwork from Reading Body Co., Reading, Pennsylvania, and was first on the run during 1957. Featuring an enclosed crew cab, two hose reels and side mounted pump, this was owned by Merchant-ville Fire Department.

Pumping 100 gallons of water a minute, this 1948 Mack is named after J Pleasants and is seen outside the fire station. A length of suction is carried at the front end of the pumper whose pump is positioned amidships. Founded by 5 brothers in 1900, the Mack Company produced its first operational apparatus in 1911, later developments including Mack ladder trucks, aircrash tenders, and aerial ladders. The Group is based in Pennsylvania and makes its own transmission systems and engines at a factory in Maryland.

Pictured at a raging house fire, the International pumper was built some 15 years ago. In addition to roof mounted beacons, the vehicle comes equipped with two red flashing lights placed above the bumper.

Built in the mid-50s, this sturdy Oshkosh machine was on the run with Oshkosh Fire Department, Wisconsin. Some interesting points include the front mounted bell and side mounted pump. More recent Oshkosh fire trucks include the massive M-23 crash/rescue vehicle carrying a 6,077 gallon (23,000 litre) water tank, plus 515 gallon (1,950 litre) foam tank.

Dramatic shot of a four wheel drive Oshkosh aircrash truck dating
from the 1940s. A front operated pump comes from Waterous of St
Paul, Minnesota, while the operator has the use of 2 foam monitors
built above the cab.

Many International fire engines are
provided with Ward LaFrance body work,
but this one has Central bodywork. Built
in 1958, this International 750/500 saw
service with Kankakee F D, Illinois.

Built some 15 years ago, the Kenworth
fire appliance shows gleaming bodywork -
a tribute to the firefighters of King
County, Washington.

An early instance of the American LaFrance 500 series, this truck was built in 1941, received power from a V8 petrol engine, and carried a 750 gallon pump. Seen in the livery of Passaic, N J, the vehicle was formerly with Middletown Township, N J, and is a good example of the 500 series fire apparatus whose 3 man seating and attractive design was in production from 1938 to 1942.

The Dennis F D, Massachusetts, has this 1967 Maxim pumper in service, capable of delivering 1000 gallons of water a minute. Much of the fire fighting gear is stowed on the outside of the vehicle whose monitor is positioned behind the cab.

Famous for its 'firsts' in fire fighting vehicles, American LaFrance introduced the first all-steel aerial ladder with full hydraulic power in 1935' and 3 years later brought out the four wheel aerials with forward cab and up to 100 foot ladder. The Angle Lake F D vehicle dates from 1971 and reflects the once popular image of 'open' fire engines with articulated rear facing ladder.